Octopuses

Illustrations: Janet Moneymaker
Design/Editing: Marjie Bassler

Copyright © 2022 by Rebecca Woodbury, Ph.D.

All rights reserved. No part of this publication may be reproduced, stored in a retrieval system, or transmitted, in any form or by any means, electronic, mechanical, photocopying, recording, or otherwise, without prior written permission from the publisher. No part of this book may be reproduced in any manner whatsoever without written permission.

Octopuses
ISBN 978-1-950415-60-1

Published by Gravitas Publications Inc.
Imprint: Real Science-4-Kids
www.gravitaspublications.com
www.realscience4kids.com

Photo credits
Cover: sandrine RONGÈRE from Pixabay; Above: Photo by Anneli Salo, CC BY SA 3.0; 1. Public Domain; 2.   Public Domain; 3. NURC/UNCW and NOAA/FGBNMS; 4. Albert Kok, CC BY SA 3.0; 5. Public Domain; 6. Rickard Zerpe, CC BY SA 2.0; 7. Daiju Azuma from KOBE, Japan, CC BY SA 2.5; 8. Vlad Tchompalov on Unsplash; 9. Katieleeosborne, CC BY SA 4.0

Octopuses are a type of **mollusk** found in the oceans.

REVIEW
MOLLUSKS

Mollusks are soft-bodied animals that live in oceans and lakes and on land.

All mollusks are in the group called **Mollusca.**

There are three different types of mollusks in the group Mollusca.

Gastropods ┄┄▶ Snails and slugs

Bivalves ┄┄▶ Clams and oysters

Cephalopods ┄┄▶ Octopuses and squids

REVIEW
MOLLUSKS

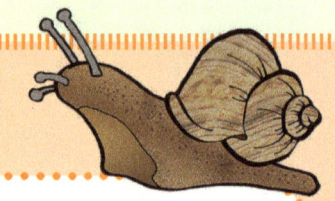

All mollusks have **organs.** An organ is a body part that has a particular job to do in the body, like the heart, stomach, and brain.

Mollusks have a **mantle** that covers the organs, and they have one or more feet or arms. Most have a shell either on the inside or outside of the body.

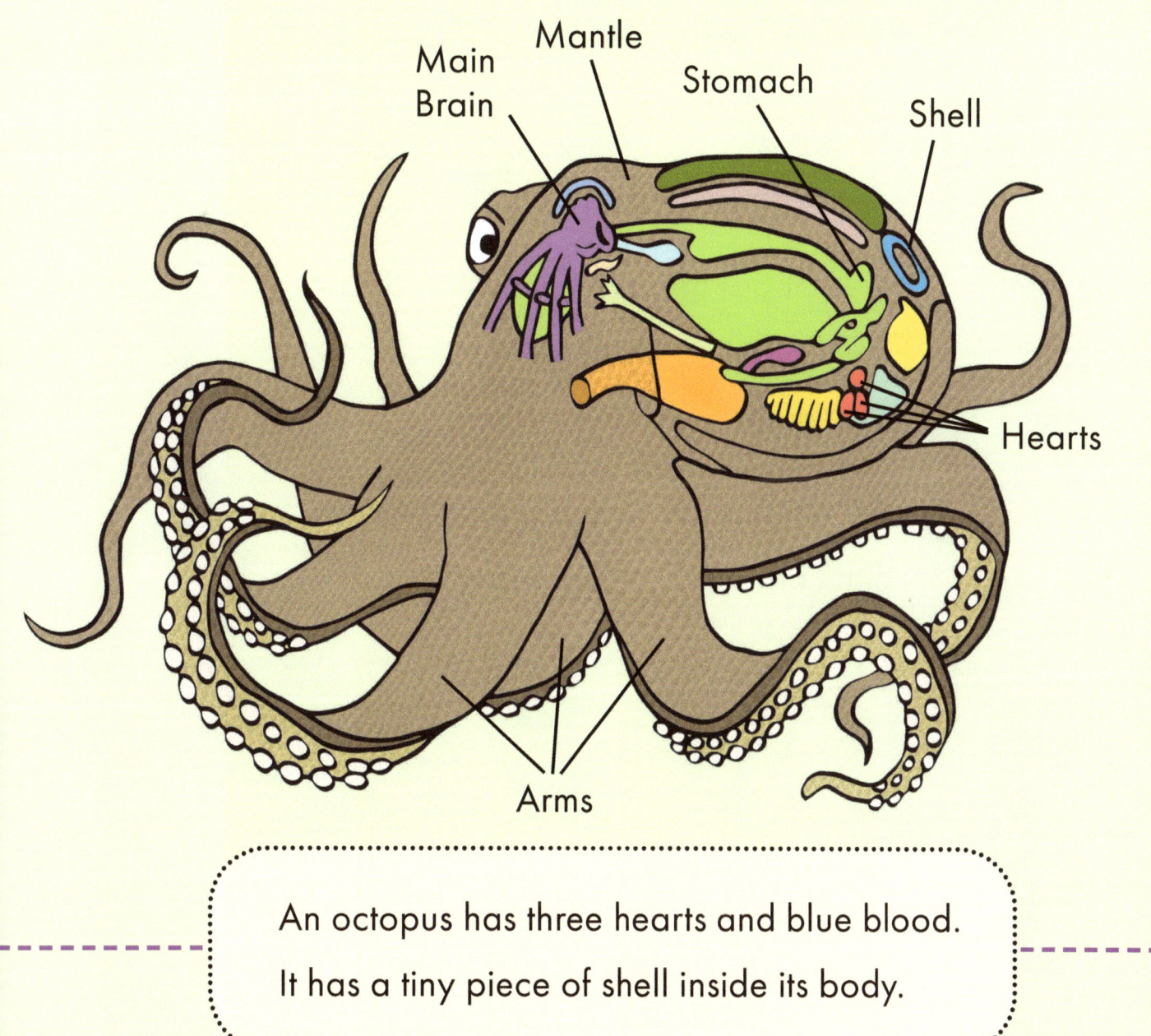

An octopus has three hearts and blue blood. It has a tiny piece of shell inside its body.

An octopus has eight arms. Its name comes from **octo**, which means eight, and **pous**, which means foot.

An octopus has nine brains. It has one main brain and a mini-brain in each arm.

Most octopuses live on the ocean floor. They move by walking with their arms.

Octopuses swim by sucking water into their body and then squirting it out behind them. This pushes them forward.

We use our arms!

Octopuses have **suckers** on their arms. They use their arms and suckers to attach themselves to objects, move objects, explore their world, and catch food.

Octopuses eat clams, crabs, snails, and fish. They can squirt out black ink to cloud the water around them and help them escape from enemies.

Ick! I eat cheese!

Octopuses have soft squishy bodies.
They can fit into small spaces.

7

Octopuses can quickly change their skin color and patterning to blend in with their surroundings.

Octopuses are very smart! They can find their way through mazes, open jars, use tools, and solve problems.

We are still learning about octopuses and what they can do.

Mice are smart too.

We can find your cheese!

How to say science words

bivalve   (BIY-valv)

brain   (BRAYN)

cephalopod   (SE-fuh-luh-pahd)

gastropod   (GAA-struh-pahd)

heart   (HAHRT)

mantle   (MAN-tuhl)

mollusk   (MAH-luhsk)

Mollusca   (mah-LUH-skuh)

octo   (AHK-toh)

octopus   (AHK-tuh-puhs)

organ   (AWR-guhn)

pous   (PUHS)

stomach   (STUH-muhk)

sucker   (SUH-kuhr)

What questions do you have about OCTOPUSES?

Learn More Real Science!

Complete science curricula from Real Science-4-Kids

Focus On Series

Unit study for elementary and middle school levels

Chemistry
Biology
Physics
Geology
Astronomy

Exploring Science Series

Graded series for levels K–8. Each book contains 4 chapters of:

Chemistry
Biology
Physics
Geology
Astronomy

www.ingramcontent.com/pod-product-compliance
Lightning Source LLC
Chambersburg PA
CBHW041632040426
42446CB00022B/3489